A Guide to Ship Repair Estimates in Man-hours

A Guide to Ship Repair Estimates in Man-hours

Don Butler

AMSTERDAM • BOSTON • HEIDELBERG • LONDON
NEW YORK • OXFORD • PARIS • SAN DIEGO
SAN FRANCISCO • SINGAPORE • SYDNEY • TOKYO

Butterworth-Heinemann is an imprint of Elsevier

Butterworth-Heinemann is an imprint of Elsevier
The Boulevard, Langford Lane, Kidlington, Oxford, OX5 1GB
225 Wyman Street, Waltham, MA 02451, USA

First edition 2003
Second edition 2012

Copyright © 2012 Don Butler. Published by Elsevier Ltd. All rights reserved.

The right of Don Butler to be identified as the author of this work has been asserted in accordance with the Copyright, Designs and Patents Act 1988

No part of this publication may be reproduced or transmitted in any form or by any means, electronic or mechanical, including photocopying, recording, or any information storage and retrieval system, without permission in writing from the publisher. Details on how to seek permission, further information about the Publisher' permissions policies and our arrangement with organizations such as the Copyright Clearance Center and the Copyright Licensing Agency, can be found at our website: www.elsevier.com/permissions

This book and the individual contributions contained in it are protected under copyright by the Publisher (other than as may be noted herein).

Notices
Knowledge and best practice in this field are constantly changing. As new research and experience broaden our understanding, changes in research methods, professional practices, or medical treatment may become necessary.

Practitioners and researchers must always rely on their own experience and knowledge in evaluating and using any information, methods, compounds, or experiments described herein. In using such information or methods they should be mindful of their own safety and the safety of others, including parties for whom they have a professional responsibility.

To the fullest extent of the law, neither the Publisher nor the authors, contributors, or editors, assume any liability for any injury and/or damage to persons or property as a matter of products liability, negligence or otherwise, or from any use or operation of any methods, products, instructions, or ideas contained in the material herein.

British Library Cataloguing in Publication Data
A catalogue record for this book is available from the British Library

Library of Congress Number: 2012935811

ISBN: 978-0-080-98262-5

For information on all Butterworth-Heinemann publications
visit our website at store.elsevier.com

Printed in the United States of America

Transferred to Digital Printing, 2012

Working together to grow
libraries in developing countries

www.elsevier.com | www.bookaid.org | www.sabre.org

ELSEVIER   BOOK AID International   Sabre Foundation

Contents

List of figures vii
List of tables ix

1 Introduction   **1**

2 Drydocking works   **5**
   Berth preparation   5
   Docking and undocking   6
   Dock rent per day   7
   Hull preparation   9
   Hull painting   11
   Rudder works   16
   Propeller works   18
   Tailshaft works   21
   Anodes   25
   Sea chests   30
   Docking plugs   30
   Valves   31
   Fenders   34

Anchors and cables	35
Chain lockers	36
Staging	37

3 Steel works — 39
Calculation of steel weights — 41

4 Pipe works — 47

5 Mechanical works — 53
Overhauling diesel engines (single-acting, slow-running, two-stroke, cross-head type) — 53
Overhauling diesel engines (single-acting, slow-running, in-line, four-stroke, trunk type) — 58
Valves — 64
Condensers — 66
Heat exchangers — 66
Turbines — 68
Compressors — 74
Receivers — 75
Pumps — 76
Boilers (main and auxiliary) — 82

6 Electrical works — 83

7 General works — 103

8 Planning charts — 107
Sample graph loadings for major trades in ship repairing — 114

Index — 119

List of figures

2.1	A vessel in dry dock sitting on keel blocks undergoing repairs	6
2.2	A vessel showing a high degree of paint damage	10
2.3	Hull preparation by water blasting (top) and hull painting by airless spray (bottom)	12
2.4	The rudder and propeller of a small vessel in dry dock	17
2.5	The rudder and propeller of a large vessel in dry dock	20
3.1	Repair of damage to shell plating	42
3.2	Repair of damage to ship's deck plating	45
4.1	Fabrication of pipe in workshop	50
5.1	A ship's medium-speed main engine	58
5.2	A ship's generator diesel engine	61
5.3	A ballast system valve chest	65
5.4	A vertical electric-driven centrifugal water pump	77

6.1	The main electrical switchboard in a machinery control room	84
6.2	A generator control panel in the main switchboard	85
6.3	A standard AC induction electrical motor	88
6.4	A ship's main diesel-driven AC alternator	92
6.5	Grouping of electric cables on a cable tray	97
6.6	Part of a distribution panel with cable attachments	100

List of tables

2.1	Shifting of blocks after docking vessel	5
2.2	Dock services	7
2.3	Removal of rudder for survey	16
2.4	Propeller works (fixed pitch) – 1	18
2.5	Propeller works (fixed pitch) – 2	19
2.6	Propeller polishing in situ (fixed pitch)	20
2.7	Tailshaft/sterntube clearances	21
2.8	Removal of tailshaft for survey	22
2.9	Gland and Simplex-type seal	24
2.10	Anodes (on hull and in sea chests)	25
2.11	Sea chests and strainers	30
2.12	Sea valves	31
2.13	Ship side storm valves	33
2.14	Hollow fenders (in half schedule 80 steel pipe)	34
2.15	Anchor cables (per side)	35
2.16	Chain lockers (per side)	36
2.17	Erection of tubular steel scaffolding, complete with all around guard rails, staging planks, and access ladders	37

3.1	Steel works renewals	43
4.1	Pipe work renewals in schedule 40 and schedule 80 seamless steel	48
4.2	Pipe clamps	51
4.3	Spool pieces	52
5.1	Top overhaul	54
5.2	Cylinder liners – 1	55
5.3	Bearing survey – 1	56
5.4	Crankshaft deflections – 1	57
5.5	Four-stroke trunk-type main engines	59
5.6	Cylinder liners – 2	60
5.7	Bearing survey – 2	62
5.8	Crankshaft deflections – 2	63
5.9	Overhauling valves, manually operated types	64
5.10	Main condenser	66
5.11	Overhauling heat exchanger	66
5.12	Main steam turbines	68
5.13	Flexible coupling	69
5.14	Auxiliary steam turbines	70
5.15	Water-tube boiler feed pumps (multi-stage type)	72
5.16	Oil tanker cargo pumps	73
5.17	Air compressor (two-stage reciprocating type)	74
5.18	Air receivers	75
5.19	Horizontal centrifugal-type pumps	76
5.20	Reciprocating-type pumps, steam driven: (a) simplex; (b) duplex	78
5.21	Reciprocating-type pumps, electric motor driven: (a) simplex; (b) duplex	79
5.22	Gear-type pumps (helical and tooth)	80

5.23	Steering gear	81
5.24	Cleaning of water-tube boilers	82
6.1	Insulation resistance tests on all main and auxiliary lighting and power circuits, and report	83
6.2	Switchboard	84
6.3	Electric motors – 1	85
6.4	Electric motors – 2	86
6.5	Electric motors for winch/windlass/crane – 1	89
6.6	Electric motors for winch/windlass/crane – 2	90
6.7	Electric generators	91
6.8	Installation of electric cables – 1	93
6.9	Installation of electric cables – 2	95
6.10	Installation of electric cables – 3	97
6.11	Installation of electric cable tray	99
6.12	Installations of electric cable conduit	101
7.1	General cleaning	103
7.2	Tank cleaning	104
7.3	Tank painting	104
7.4	Tank testing	105

1 Introduction

This guide has been produced in order to outline, to technical superintendents of shipowners and ship managers, the manner in which the commercial departments of ship repairers compile quotations. The ship repairers use their tariffs for standard jobs to build up their quotations. This guide is based on these tariffs, but is made up in man-hours to assist long-term pricing. It can also be of assistance to shipyards without this information to prepare man-hour planning charts, helping them to assess manpower requirements for jobs and to produce time-based plans. Man-hours have been used so that this book will not be 'dated' and can be used without encountering the problems of increases in costs over the years. Where man-hour costs are not possible, these have been noted and suggestions made to compile costs against these items.

It is to be noted that, apart from steel works and pipe works, no cost of materials has been included within this book. Only man-hours are used in order that the compiler may assess shipyards' charges based on the current market price of labor.

Where materials are conventionally supplied by the repair contractor, these have been built in to the labor costs and evaluated as man-hours. Apart from steel works and pipe works, the cost of materials in the jobs listed is generally minimal when compared with labor costs. So, apart from these two, most of the other costs will be consumables.

A comparison between various countries has been included. The workers of some countries have more efficient skills than do others. Some establishments have more sophisticated equipment than others. However, common ground has been assumed in the output of workers in standard jobs.

It is stressed that this book considers only ship 'repairs'. That is, removing damaged, worn, or corroded items, making or supplying new parts to the pattern of the old, and installing. It is not meant to be used in its entirety for new building work, although, in some areas, it may prove useful.

Unless specifically mentioned, all the repairs are 'in situ'. For removing a specific item ashore to the workshops, consideration should be given to any removals necessary to facilitate transportation through the ship and to the shore workshop and the later refitting of these removals, and an appropriate charge made.

In calculating the labor man-hours, it should be borne in mind that these will vary for similar jobs carried out under different conditions, such as world location, working conditions, environment, type of labor, availability of backup labor, etc.

The labor times given in this book are based upon the use of trained and skilled personnel, working in reasonable conditions in an environment of a good quality ship repair yard with all necessary tools, equipment, and readily available materials and consumables.

All of these factors should be considered when calculating the man-hours and if conditions vary from those of the assumption of this book then factors should be applied to compensate for any shortfall in any conditions. As an example, if the work is being carried out in a country that suffers from heat and high humidity, then the output of a worker can fall to 50% that of the same worker in another country that has an easier working climate.

With reduced work outputs for whatever reason, a ship repair yard will need to mark up their pricing rates according to their type of variance, and this is passed on to the shipowner. The estimator should consider influences applicable and may need to apply a factor to increase the man-hours according to whatever may reduce the output of a contractor's workers.

Once the man-hours have been calculated, the estimator must then apply a pricing rate to the total. These vary from place to place and should be ascertained from the ship repair establishments under consideration. The variance of the rates will be applicable to certain considerations that can be applied. These considerations can include local economy, how hungry the yard is for work, current workload of the yard, and other similar situations. The estimator can look at the economical climate of the repair yards and ascertain a variance factor for each yard and apply these accordingly.

The figures shown in this book are not to be viewed as invariable. Obviously different shipyards have different working conditions and techniques, so the man-hours for the work can vary. However, the figures shown can be used as a fair assessment of the work in general and can produce price estimates for budget purposes to a shipowner. This is the object of the book.

When requesting quotations from shipyards the quotes received always vary tremendously. The figures given in this book reflect competitive tariff rates.

The author has long-term experience in the ship repair world, having recently retired as a director of a marine consultancy and is now running his own consultancy, albeit on a part time basis. An ex-sea-going engineer, qualified and experienced in steam and motor ships, he even has experience of steam reciprocating engines and saturated steam fire tube boilers. From there, he rose to repair superintendent. He has extensive ship repair yard experience gained from production, commercial, and general manager positions.

Seeing a lack of this type of publication, the author decided to put his long-term experience to use in order to assist those responsible for compiling repair specifications with pricing strategy so they may build up costings for their planned repair periods.

Included in the text are a number of tips to be applied in the preparation of repair specifications and finalizing contracts with ship repair yards. The wording of much of the scope of works listed in the book may be used within a repair specification, so as to clearly outline the owners' requirements.

2 Drydocking works

Berth preparation

This item is included within the charges for docking and undocking, and should also include those for dismantling and removal of any specially prepared blocks.

Table 2.1 Shifting of blocks after docking vessel

This covers shifting of blocks at the request of the owner for access works not known at the time of quoting. This involves cutting out the soft wood capping of the block, shifting the block, and reinstalling at a different location.

DWT	Keel block man-hours	Side block man-hours
<20,000	5	3
20,000–100,000	10	5
100,000–200,000	16	8
>200,000	20	12

Docking and undocking

This is a variable dependent upon world location and market demands. Drydocking charges regularly change depending upon the economical climate, so an owner's superintendent should check with selected drydock owners for their current rates.

Figure 2.1 A vessel in dry dock sitting on keel blocks undergoing repairs

Dock rent per day

The above comments also apply here.

Table 2.2 Dock services

Service	<100 LOA man-hours	>100 LOA man-hours
Fire and Safety watchman per day	8/shift	8/shift
Garbage skip per day	2	4
Electrical shore power connection and disconnection	4	5
Electrical shore power per unit	Variable	Variable
Temporary connection of fire main to ship's system	5	6
Maintaining pressure to ship's fire main per day	3	3
Sea circulating water connection	3	4
Sea circulating water per day	4	4
Telephone connection on board ship	3	3
Supply of ballast water per connection	6	8

Continued

Table 2.2 Dock services—cont'd

Service	<100 LOA man-hours	>100 LOA man-hours
Supply of fresh water per connection	3	5
Connection and disconnection of compressed air	3	5
Gas-free testing per test/visit and issue of gas-free certificate	8	10
Electric heating lamps per connection	4	5
Ventilation fans and portable ducting each	5	5
Wharfage: charges to lie vessel alongside contractor's berth, usually a fixed rate per meter of vessel's length	Variable	Variable
Cranage: charges variable, dependent upon size of crane	Variable	Variable

Notes:
Contractors often charge for temporary lights provided for their own use in order to carry out repairs. This is an arguable point as it is for their benefit and not the owner's. It should be classed as an overhead and costed accordingly. Provided there are none of the ship's staff utilizing the temporary lights, then it should be a contractor's cost and included within the original quotation.

Hull preparation

- Hand-scraping normal
- Hand-scraping hard
- Degreasing before preparation works
- High-pressure jet wash (up to 3000 p.s.i.)
- Water blast
- Vacuum dry blast
- Dry blast (dependent upon world location, prohibited in some countries)
- Grit sweep
- Grit blast to Sa2
- Grit blast to Sa2.5
- Spot blast to Sa2.5
- Hose down with fresh water after dry blast
- Disk preparation to St2.

The charges for hull preparation works should be given in price per square meter. This will enable the owner's superintendent to calculate the price for the full scope of works.

Special notes for hull preparation

The shipowner's superintendent should be fully aware of the manner in which the ship repair yard has quoted for the hull preparation works. This is to obviate surprise items when confronted with the final invoice.

Figure 2.2 A vessel showing a high degree of paint damage

A ship repair yard should quote fully inclusive rates, which cover the supply of all workers, equipment, machines, tools and consumables to carry out the quoted works, and also for all final cleaning-up operations. Inflated invoices have been known from shipyards covering the removal of used blasting grit, removed sea growth, etc. The dry dock may not belong to the repair contractor and additional charges may be made by the dry dock owner for these items. Ensure that these charges are well highlighted before acceptance of the quotation. It is far better to clear up these matters prior to the arrival of the vessel instead of being involved in arguments just before the vessel sails. Time taken to consider what a yard may see as justifiable extras before the event is well spent prior to placing the order, when everyone in the yard is eager to secure the contract.

The use of dry blasting grit is being phased out in certain areas due to its being environmentally unfriendly. Dry sand is not used for similar reasons and is also a health hazard.

In these cases the choice is for vacuum dry blasting or water blasting using very high pressures or wet blasting using grit in the water stream. Water blasting can use fresh or salt water, but the salt water cleaning must be followed by thorough high-pressure jet washing using fresh water to remove the salts.

The shipowner must determine the blasting method that is to be used by the shipyard in removing the old paint from the hull plating and obtain their fully inclusive quotation for this work. This book does not give man-hour rates for hull cleaning, as yards generally quote per square meter. Within the book a method is shown on how to calculate the square meter area of the hull of a vessel, so this should be used in conjunction with the quoted rate per square meter to determine the final cost of this work.

Hull painting

- Flat bottom
- Vertical sides
- Topsides
- Touch-up after spot blast
- Names, homeport, load lines, draft marks.

The charges for hull painting works should be given in price per square meter, and a fixed rate for names and marks. This will enable the owner's superintendent to calculate the full price for the scope of works (see below for method of determining painting areas of a ship's hull).

Figure 2.3 Hull preparation by water blasting (top) and hull painting by airless spray (bottom)

Notes for hull painting

Shipyard standard rates will apply for paints considered as 'normal'. This refers to paints being applied by the airless spray method up to a maximum of 100 microns (μ) dry film thickness (dft) and having a drying time between applications not exceeding 4 hours. The owner should ensure that the shipyard is aware of any special, or unconventional, painting compositions that may be used. If this is not highlighted in the specification, the contractor is justified in claiming extra costs.

Additional note on the supply of painting compositions

It is generally accepted practice for all painting compositions to be the owner's supply. This is due to the paint manufacturer giving their guarantee to the *purchaser* of their paints. Included from the manufacturer, within the price of the paints, is their technical backup, provision of a technical specification on the preparation works and paint application, and the provision of a technical supervisor to oversee the whole process of the paint application. If the paints have been applied to the satisfaction of the technical representative, then the full guarantee will be given to the purchaser by the paint manufacturer.

The contractor is only responsible for the preparation works and the application of the painting compositions. Provided they have satisfied the conditions of the technical specification, and the attending technical representative, then there will be no comeback on them if a problem with the paints occurs at a later date.

With the owner being the purchaser, the paint manufacturer will have the responsibility to provide new paint in the event of problems. The application is the responsibility of the owner. He will have to bear the cost of drydocking the ship and having the replacement paints applied.

If the ship repair contractor supplies the paints, he will be responsible for all these costs incurred. Hence it is not in the interests of the ship repair contractor to supply the painting compositions.

Formula to determine the painting area of ship hulls

Input the following data:

LOA in meters	xxx
LPP in meters	xxx
BM in meters	xx
Draft max. in meters	xx
P = UW constant for type of hull (0.7 for fine hulls, 0.9 for tankers)	0.x
Height of boot-top in meters	xx
Height of topsides in meters	xx
N = constant for topsides for type of hull (0.84–0.92)	0.xx
Height of bulwarks in meters	xx

Underwater area including boot-top	
Boot-top area	
Topsides area	
Bulwarks area	

Underwater area including boot-top

$$\text{Area} = \{(2 \times \text{Draft}) + \text{BM})\} \times \text{LPP} \times \text{P(constant for vessel shape)}$$

Boot-top area

$$\text{Area} = \{(0.5 \times \text{BM}) + \text{LPP}\} \times 2 \times \text{Height of boot-top}$$

Topsides area

$$\text{Area} = \{\text{LOA} + (0.5 \times \text{BM})\} \times 2 \times \text{Height of topsides}$$

Bulwarks area (note: external area only)

$$\text{Area} = \{\text{LOA} + (0.5 \times \text{BM})\} \times 2 \times \text{Height of bulwarks}$$

Using the above formulae, it is a simple matter to formulate a computer spreadsheet to determine the external painting areas of the vessel. Input the data into the table and use the formulae to determine the external painting areas of the vessel.

Rudder works

Table 2.3 Removal of rudder for survey

(a) Repacking stock gland with owner's supplied packing. Measuring clearances, in situ.
(b) Disconnecting rudder from palm and landing in dock bottom for survey and full calibrations. Refitting as before on completion.

| | Man-hours | |
DWT	(a)	(b)
>3000	15	165
5000	18	250
10,000	20	280
15,000	25	300
20,000	28	350
30,000	30	400
50,000	35	500
80,000	45	600
100,000	60	800
150,000	75	900
200,000	90	1000
250,000	110	1200
350,000	120	1500

Figure 2.4 The rudder and propeller of a small vessel in dry dock

Propeller works

Table 2.4 Propeller works (fixed pitch) – 1

(a) Disconnecting and removing propeller cone, removing propeller nut, setting up ship's withdrawing gear, rigging and withdrawing propeller, and landing in dock bottom. On completion, rigging and refitting propeller as before and tightening to instructions of owner's representative. Excluding all removals for access, any other work on propeller, and assuming no rudder works.

(b) Transporting propeller to workshops for further works and returning to dock bottom on completion.

	Man-hours	
Shaft dia. (mm)	(a)	(b)
Up to 100	20	15
100–200	30	18
200–300	45	25
300–400	60	30
400–800	90	60
800–900	150	100

Table 2.5 Propeller works (fixed pitch) – 2

(a) Receiving bronze propeller in workshop, setting up on calibration stand, cleaning for examination, measuring and recording full set of pitch readings. Polishing propeller, setting up on static balancing machine, checking and correcting minor imbalances.
(b) Heating, fairing, building up small amounts of fractures and missing sections, grinding and polishing.

	Man-hours	
Dia. (mm)	(a) Manganese bronze	(b) Aluminum bronze
Up to 400	15	21
400–800	32	42
800–1200	52	68
1200–1800	75	85
1800–2000	90	105
2000–2500	100	125
2500–3000	130	150
3000–4000	150	180
4000–5000	180	210

Note: Covers repairs outside 0.4 blade radius only; classed as minor repair.

Figure 2.5 The rudder and propeller of a large vessel in dry dock

Table 2.6 Propeller polishing in situ (fixed pitch)

Polishing in situ using high-speed disk grinder, coating with oil; ship in dry dock.

Dia. (mm)	Man-hours
Up to 400	6
400–800	11
800–1200	17
1200–1800	25
1800–2000	28
2000–2500	35
2500–3000	50
3000–4000	80
4000–5000	120

Tailshaft works

Table 2.7 Tailshaft/sterntube clearances

Removing rope-guard, measuring and recording wear-down of tailshaft and refitting rope-guard, including erection of staging for access, by:
(a) Feeler gauge.
(b) Poker gage coupled with jacking up shaft.
(c) Repacking internal sterngland using owner's supplied soft greasy packing.

Tailshaft dia.	Man-hours		
(mm)	(a)	(b)	(c)
Up to 150	10	15	7
150–250	15	22	11
250–300	21	30	14
300–400	30	40	30
400–800	35	45	35
800–1200	50	55	—
1200–1800	—	57	—
1800–2000	—	60	—

Drydocking works

Survey

Table 2.8 Removal of tailshaft for survey

Disconnecting and removing fixed pitch propeller and landing in dock bottom.

(a) Disconnecting and removing tapered, keyed, inboard tailshaft coupling, drawing tailshaft outboard and landing in dock bottom for survey, cleaning, calibrating, and refitting all on completion.

(b) Disconnecting inboard intermediate shaft fixed, flanged couplings, releasing the holding down bolts of one in number journal bearing, rigging intermediate shaft, lifting clear and placing in temporary storage on ship's side. Assuming storage space available. Withdrawing tailshaft inboard, hanging in accessible position, cleaning, calibrating, and refitting on completion. Relocating intermediate shaft and journal bearing in original position, fitting all holding-down bolts and recoupling flanges all as before.

Includes erection of staging for access.

Includes repacking inboard gland using owner's supplied, conventional soft greasy packing.

Excludes any repairs.

Excludes any work on patent gland seals.

Tailshaft dia. (mm)	Man-hours	
	(a) Withdrawing tailshaft outboard	(b) Withdrawing tailshaft inboard
Up to 150	90	140
150–250	120	180
250–300	200	250
300–400	300	400
400–800	500	600
800–1200	—	1000
1200–1800	—	1200

Crack detection

- Magnaflux testing of tailshaft taper and key way.
- Allowance made of 8 man-hours for the testing works, which is performed after all removals for access.

Table 2.9 Gland and Simplex-type seal

(a) Removing gland follower, removing existing packing from internal stern gland, cleaning out stuffing box, and repacking gland using owner's supplied conventional soft greasy packing.
(b) Disconnecting and removing forward and aft patent mechanical seals (Simplex type). Removing ashore to workshop, fully opening up, cleaning for examination and calibration. Reassembling with new rubber seals, owner's supply.

Excluding all machining works.
Assuming previous withdrawing of tailshaft.

	Man-hours	
Tailshaft dia. (mm)	(a)	(b)
Up to 150	8	—
150–250	12	—
250–300	15	35
300–400	23	50
400–800	30	110
800–1200	35	150
1200–1800	—	200
1800–2000	—	230

Anodes

Table 2.10 Anodes (on hull and in sea chests)

Cutting off existing corroded anode, renewing owner's supplied zinc, or aluminum, anode by welding integral steel strip to ship's hull. Excluding all access works.

Weight (kg)	Man-hours
3	1
5	1
10	1.5
20	2

To determine the amount of anodes required for a vessel, the owner should contact a supplier who will calculate the exact requirement.

The following section shows the method of determining weights of zinc and aluminum anodes, so the reader may understand the method. (See also the section on hull painting for method of determining underwater area of ship's hulls, which is required in order to determine the amount of anodes required for a vessel.)

Formula to determine the weight of sacrificial zinc anodes required on a ship's underwater area

Underwater area of ship in square meters	xxx,xxx
Number of years between anode change	3
Capacity of material in amp hours/kg	781
Current density of material in mA/m^2 (ave. 10–30)	20
K	8760

Formula for total weight of sacrificial zinc anodes (kg)

$$= \frac{\text{Current (amps)} \times \text{Design life (years)} \times \text{K (8760)}}{\text{Capacity of material (amp hours/kg)}}$$

where:

Current (amps)
$$= \frac{\text{Underwater area (m}^2) \times \text{Current density}}{1000}$$

Current density of material in mA/m^2
= Information from manufacturer
(between 10 and 30, say 20)

Design life
= Number of years between dry dockings (e.g. 3)

K = Number of hours in 1 year = 8760

Capacity of material (amp hours/kg)
= Information from manufacturer
(for zinc, 781 is common)

Using the above formula, it is a simple matter to formulate a spreadsheet to determine the weight of zinc anodes.

Input the data into the table and use the formula to determine the weight of zinc anodes for the period required.

Formula to determine the weight of sacrificial aluminum anodes required on a ship's underwater area

Underwater area of ship in sq. meters	xxx,xxx
Number of years between anode change	3
Capacity of material in amp hours/kg	2600
Current density of material in mA/m^2 (ave. 10–30)	20
K	8760

Formula for total weight of sacrificial zinc anodes (kg)
$$= \frac{\text{Current (amps)} \times \text{Design life (years)} \times K\ (8760)}{\text{Capacity of material (amp hours/kg)}}$$

where:

Current (amps)
$$= \frac{\text{Underwater area (m}^2) \times \text{Current density}}{1000}$$

Current density of material in mA/m^2
= Information from manufacturer
(between 10 and 30, say 20)

Design life
 = Number of years between dry dockings (e.g. 3)

K = Number of hours in 1 year = 8760

Capacity of material (amp hours / kg)
 = Information from manufacturer
   (for aluminum, 2600 is common)

Using the above formula, it is a simple matter to formulate a spreadsheet to determine the weight of zinc anodes.

Input the data into the table and use the formula to determine the weight of aluminum anodes for the period required.

From the above tables and results, it will be seen that approximately three to four times the weight of zinc anodes are required, to that of aluminum, for the same protection. It is for this reason that, in larger vessels, aluminum is used in preference to zinc. Of course, aluminum is much more expensive than zinc.

Sea chests

Table 2.11 Sea chests and strainers

Opening up of sea chests by removing ship side strainers, cleaning, and painting with owner's paints, as per hull treatment specification. Assuming single grid per chest.

Surface area (m^2)	Man-hours
Below 0.3	12
0.3–1	20
Above 1	30
Additional charge per extra grid	5

Docking plugs

Allowance made of 1 man-hour for removing and later refitting of each tank drain plug using ship's spanner, assuming no locking devices fitted and excluding all removals for access and repairs to threads.

Valves

Table 2.12 Sea valves

Opening up hand-operated, globe and gate valve for in-situ overhaul, by disconnecting and removing cover, spindle and gland, cleaning all exposed parts, hand grinding of globe valve, light hand-scraping of gate valve, testing bedding, painting internal exposed areas, and reassembling with new cover joint and repacking gland with conventional soft packing.

(a) Butterfly valve, remove, clean, check, testing bedding of seals, paint internal exposed areas and refit; excluding operating gear.
(b) Checking and cleaning large butterfly valves through the sea chest.

Valve bore (mm)	Globe valve	Gate valve	(a) Butterfly valve	(b) Butterfly valve
>50	4	4.5	6	—
100	6	7	8.5	—
150	8	9	11.5	—
200	10	11	14	—
250	13	14	18	—
300	16	17	22	—
350	20	21	26	13
400	23	24	29	14
450	26	28	33	14.5
500	30	31	37	15
550	34	35	42	16

Continued

Table 2.12 Sea valves—cont'd

Valve bore (mm)	Globe valve	Gate valve	(a) Butterfly valve	(b) Butterfly valve
600	37	39	46	16.5
650	41	43	51	17
700	44	47	56	18
750	47	49	60	19
800	50	53	66	20
900	57	60	81	22
1000	65	68	100	24
1100	73	77	106	25
1200	84	88	113	27
1300	95	100	120	30

Notes:
Valves in pump rooms, additional 15%.
Valve in cofferdams and inside tanks, additional 20%.
Removals for access not included.
Staging for access not included.
Removing valve ashore to workshop for above type of overhaul requires special consideration, dependent upon size. Valves below 20 kg in weight can be assessed as double the in-situ rate. Above this requires rigging and cranage input, which should be assessed separately.

Table 2.13 Ship side storm valves

Opening up ship side storm valve for in-situ overhaul, by disconnecting and removing cover, spindle and gland, cleaning all exposed parts, testing bedding, painting internal exposed areas and reassembling with new cover joint and repacking external gland with conventional soft packing.

Dia. (mm)	Man-hours per valve
50	9
75	10
100	12
125	14
150	16
200	17

Note: Disconnecting and removing ashore for above overhaul and later refitment; double the above rate.

Fenders

Table 2.14 Hollow fenders (in half schedule 80 steel pipe)

Fendering formed by cutting steel pipe into two halves.
Cropping existing external damaged fendering, hand-grinding remaining edges and preparing remaining flat hull plating for welding.
Supplying and fitting new fendering in half round standard schedule 80 steel pipe and full fillet welding fender in place.
Including erection of staging for access and later dismantling.
Exclusions: All hull preparation and painting of the steel works in way of the repairs.

	Man-hours per meter	
Pipe dia. (mm)	Straight run of fender	Curved fender at corners
200	20	30
250	22	32
300	24	34
350	26	36

Note: The above figures are for split steel pipe only. For other shapes, steel fabrication tariffs will be applicable, based upon steel weights.

Anchors and cables

Table 2.15 Anchor cables (per side)

Ranging out for examination and later restowing.
Cleaning by high-pressure jet wash or grit sweeping.
Calibration of every 20th link and recording.
Marking shots with white paint.
Painting cables with owner's supplied bitumastic paint.
Opening 'Kenter' type shackle and later closing.
Disconnect first length of cable and transferring to end.
Changing cable end for end.

Small vessels

Cable dia. (mm)	Man-hours (per side)
<25	70
25–50	90

Large cargo vessels and oil tankers

DWT	Man-hours (per side)
<20,000	100
20,000–50,000	130
50,000–100,000	140
100,000–200,000	200
200,000–300,000	250
Over 300,000	270

Chain lockers

Table 2.16 Chain lockers (per side)

Opening up, removing dry dirt and debris, handscaling, cleaning, and painting one coat bitumastic. Closing up on completion.
Removing internal floor plates, or grating, cleaning, painting, and refitting.

Small vessels

Cable dia. (mm)	Man-hours (per side)
<25	75
25–50	90

Large cargo vessels and oil tankers

DWT	Man-hours (per side)
<20,000	100
20,000–50,000	130
50,000–100,000	140
100,000–200,000	200
200,000–300,000	250
Over 300,000	270

Note: Removal of sludge will be charged extra per m^3.

Staging

This item is usually charged within a particular job. When included within the charge of a job, that job price is increased accordingly. However, to assist estimating, it can be based on cubic meters of air space covered. A minimum charge of approximately 8 m^3 will be made.

The figures stated in Table 2.17 cover erection and later dismantling and removal of external staging. For internal staging, inside tanks, engine rooms, etc., a third column is shown.

Table 2.17 Erection of tubular steel scaffolding, complete with all around guard rails, staging planks, and access ladders

	Man-hours/m^3	
m^3	External	Internal
Up to 10	3	5
10–100	2.5	4
>100	2	3

3 Steel works

Applicable to Grade A shipbuilding steels.

- Marking off the external area of hull plating on vertical side up to a height of 2 meters, cropping by hand burning, and removal of all cropped plating.
- Dressing and preparation of plate edges of remaining external plating.
- Dressing and preparation of remaining internal structure.
- Supply and preparation of new flat steel plating, blasting to Sa2.5, and applying one coat of owner's supplied holding primer.
- Transportation of new plate to vessel, fitting up, wedging in position, minor fairing and dressing of plate edges in the *immediate* vicinity, applying first runs of welding on one side, back gouging from other side, and finally filling and capping to give fully finished weld.

Included in the tariff are:

- Only the work to the steel work mentioned.
- Cleaning and chipping paint in the *immediate* vicinity of the repair area to facilitate hot cutting work.
- Cranage and transportation of the new and removed steelwork.

Exclusions are:

- Staging for access. For staging charges see relevant section.
- All removals for access and later refitments.
- Tank cleaning and gas freeing.
- Cleaning in way of repairs other than the immediate vicinity as noted above.
- All final tests to repairs.
- Fairing of adjacent plates except as minor in the immediate vicinity as noted above.

Man-hours are per tonne of finished dimensions. The rates shown are for large quantities of steel renewals. The lower limit will be given by the shipyard and is dependent upon the size of the repair yard and the vessel. Assume the limit to be approximately 5 metric tonnes.

Shipowner superintendents should be aware of the methods used by the shipyards of calculating steel weights, and this is illustrated below.

Calculation of steel weights

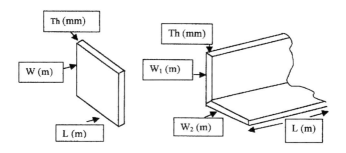

FLAT PLATE
$L \times W \times Th \times SG$
= Weight in kg

ANGLE
$L \times (W_1 + W_2) \times Th \times SG$
= Weight in kg

Flat steel plate

For flat steel plates, measure the length in meters, the width in meters, and the plate thickness in millimeters. Take the specific gravity (SG) of the material. For steel, the SG is 7.84, but it is common practice for estimators to use 8. To calculate the weight of the plate in kilograms:

$$\text{Multiply } L \times W \times Th \times SG$$

For example:

Plate no.	L (m)	W (m)	Th (mm)	SG	Weight (kg)
1234	1	1	10	8	80

Steel works

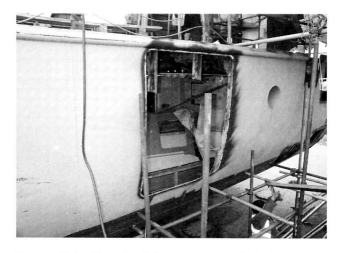

Figure 3.1 Repair of damage to shell plating

Steel angle

For flat steel angles, measure the length in meters, the widths of each leg in meters, and common thickness in millimeters. Take the specific gravity (SG) of the material. For steel, the SG is 7.84, but it is common practice for estimators to use 8. To calculate the weight of the steel angle in kilograms:

$$\text{Multiply } L \times (W_1 + W_2) \times Th \times SG$$

For example:

Angle no.	L (m)	W_1 (m)	W_2 (m)	Th (mm)	SG	Weight (kg)
1111	1	0.150	0.175	10	8	26

For other steel sections, break each down into separate flat sections, calculate individually, and finally add together to obtain the total weight of the section.

Table 3.1 Steel works renewals

Plate thickness (mm)	Man-hours per tonne
Up to 6	250
8	245
10	240
12.5	230
16	220
18	210
20	200

Correction for curvature	Factor increase
Single	1.2
Double	1.3

Correction for location – **external**	Factor increase
Flat vertical side above 2 meters in height and requiring staging for access	1.1
Bottom shell, accessible areas (i.e. no removals of keel blocks)	1.12

Continued

Table 3.1 Steel works renewals—cont'd

Correction for location – **external**	Factor increase
Keel plate	1.4
Garboard plate	1.25
Bilge strake	1.25
Deck plating	1.15

Correction for location – **internal**	Factor increase
Bulkhead	1.2
Longitudinal/transverse above DB areas	1.25
Longitudinal/transverse below DB areas	1.35

Other adjustment factors	Man-hour adjustment
For fairing works:	
Remove, fair, and refit	80% of renewal price
Fair in place (if practicable)	50% of renewal price

Note: For high tensile grade AH shipbuilding steels, increase rates by 10%.

Figure 3.2 Repair of damage to ship's deck plating

Notes for steel works renewals

- The steel weight is calculated from the maximum dimensions of each single plate and applying a specific gravity of 8.
- Staging for access and cranage is normally included within the price differences for repair locations. This should be checked with the contractor.

- A minimum quantity of steel renewals per area is usually stated in a ship repairer's tariff or conditions and is dependent upon the size of the shipyard and the size of the vessel. Below this minimum weight per area, the repairer will either charge anything up to double the standard tariff or will charge labor time and materials.
- If a plate is being joined to an adjacent plate of different thickness, then an additional labor charge will be made for taper grinding of the thicker plate to suit.

4 Pipe works

Table 4.1 Pipe work renewals in schedule 40 and schedule 80 seamless steel

Removal of existing pipe and disposal ashore. Fabrication of new pipe in workshop to pattern of existing complete with new flanges, delivery on board of new pipe, and installation in place with the supply of new soft jointing and standard material fastenings. Refitment of original clamps with new standard material fastenings.

Inclusions:
- Pipes in straight lengths, no branches, and with two flanges. Up to 50 mm nominal bore pipes can be screwed. Above 50 mm nominal bore pipes can be supplied with slip on welded flanges.
- Pipes readily accessible on deck or in engine rooms above floor plate level.

Exclusions:
- Access works.
- Removals for access. This also includes other pipes in way.
- Any cleaning works.
- Heating coils. These are subject to special consideration.
- Any necessary staging or access works.

Continued

Table 4.1 Pipe work renewals in schedule 40 and schedule 80 seamless steel—cont'd

Pipe dia. (inches)	Man-hours per meter	
	Schedule 40 steel straight pipe	Schedule 80 steel straight pipe
>0.5	2	2.3
1	2.5	2.6
1.5	3.2	3.9
2	4.2	5
2.5	5.2	6.5
3	6.3	8
4	8.4	10
5	10.5	13
6	12.5	16
8	16.5	21
10	21	26
12	25	31
14	29	37
16	34	42
18	38	47
20	42	53
22	46	58
24	50	64

Notes:
A minimum charge to be applied for length of pipe of 3 meters.
Per bend add 30% of the value of the straight pipe.
Per branch add 80% of the value of the straight pipe.
Removal and refitment only of the pipe: charge is 40% of the value of the pipe.

For pipes in other locations, the following additional charges are to be made:

Inside double bottom tanks and duct keels	30%
Inside cargo or ballast tanks	30%
Inside pump rooms	30%
In engine rooms below floor plate level	20%

Galvanizing

Hot dip galvanizing after manufacture.
15% of finished steel pipes.

Ready galvanized pipes

20% of finished steel pipes.

Copper pipes

Pipe work renewals in copper.
The rate for copper pipe renewals is estimated as 300% that of schedule 40 steel pipes.

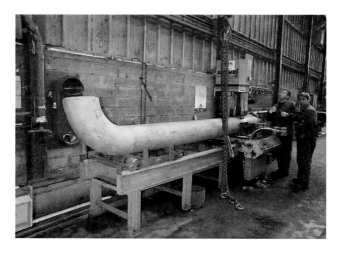

Figure 4.1 Fabrication of pipe in workshop

Table 4.2 Pipe clamps

Supply and fitting of new pipe clamps together with the supply of new standard material fastenings. Including welding of clamp to ship's structure.

Pipe dia. (inches)	Man-hours per renewal of pipe clamp
>3	2
4	2.5
5	3
6	3.5
8	4
10	5
12	6
14	8
16	9
18	11
20	12
22	13
24	14

Table 4.3 Spool pieces

Removal of existing steel penetration pieces from bulkheads or decks, fabrication and installation of new straight, seamless steel, penetration pieces with two flanges and one flat bulkhead compensation flange welded in place. Including supply and installation of soft joint and standard material fastenings.

	Man-hours per spool piece	
Pipe dia. (inches)	Schedule 40 steel straight pipe	Schedule 80 steel straight pipe
>0.5	2	2.5
1	3.5	4.3
1.5	5.5	6.8
2	9	11.2
2.5	10	12.5
3	12.5	15.5
4	15	18.7
5	17	21
6	19	24
8	29	36
10	35	44
12	42	52
14	51	64
16	62	77
18	73	91
20	85	105
24	95	118

Note: The same conditions apply to these spool pieces as for pipe renewals.

5 Mechanical works

In the following mechanical works it is assumed that all are in-situ overhauls and that the items considered are all accessible for the work to be performed.

If any item is to be removed ashore to the workshops then an assessment of the work involved must be made and the rates given amended accordingly. This is considering removals to permit transportation of the items from the ship and their later refitment, on completion of the reinstallation.

Overhauling diesel engines (single-acting, slow-running, two-stroke, cross-head type)

In the overhauling of large main propulsion engines, it is assumed that the ship will provide all the specialized equipment necessary. This refers to heavy-duty equipment that is normally supplied by the engine manufacturer and also any special hydraulic tensioning equipment for fastenings. Regarding main engines, it is assumed that there are lifting devices such as overhead beams, runner blocks, and/or overhead cranes.

Table 5.1 Top overhaul

Disconnect and remove cylinder head, withdraw piston, remove piston rings, clean, calibrate, and reassemble as before using all owner's supplied spares.

Cylinder bore (mm)	Man-hours per cylinder
500	67
550	70
600	75
650	80
700	87
750	95
800	105
850	115
900	125
950	137
1000	150

Cylinder cover

Disconnect and remove cylinder head, clean all exposed parts, including piston crown, and reassemble as before using all owner's supplied spares.

Assume as 60% the rate of top overhaul.

Table 5.2 Cylinder liners – 1

Withdrawal of cylinder liner, cleaning exposed areas as far as accessible, installation of new, owner's supplied liner and rubber seals.

Cylinder bore (mm)	Man-hours per cylinder
500	60
550	62
600	67
650	72
700	77
750	82
800	90
850	97
900	107
950	120
1000	135

Note: Assuming that cylinder head, piston, and piston rod are already removed as part of the top or complete overhaul.

Table 5.3 Bearing survey – 1

Opening up for inspection, exposing both halves, cleaning, calibrating, and presenting for survey. On completion, reassembling as before.

	Man-hours/bearing		
Cylinder bore (mm)	*Cross-head*	*Crank pin*	*Main*
500	47	32	45
550	50	35	48
600	52	37	50
650	55	40	52
700	57	42	55
750	58	45	56
800	60	47	59
850	62	50	63
900	65	52	67
950	68	55	72
1000	72	57	76
1050	75	60	80

Note: For exposing top half of bearing shell only, charge 60% of above rates.

Table 5.4 Crankshaft deflections – 1

(a) Opening up crankcase door for access works and refitting on completion.
(b) Setting up deflection indicator gage, turning engine, using ship's powered turning gear, and recording observed readings. Removing equipment and closing up crankcase door on completion.

Cylinder bore (mm)	(a) Deflections (man-hours per unit)	(b) Crankcase doors (man-hours per door)
>750	4	4
750–850	4.5	4.5
850–950	4.5	5
950–1050	5	5.5

Overhauling diesel engines (single-acting, slow-running, in-line, four-stroke, trunk type)

In the overhauling of large main propulsion engines, it is assumed that the ship will provide all the specialized equipment necessary. This refers to heavy-duty equipment that is normally supplied by the engine manufacturer and also any special hydraulic tensioning equipment for fastenings. Regarding main engines, it is assumed that there are lifting devices such as overhead beams, runner blocks, and/or overhead cranes.

Figure 5.1 A ship's medium-speed main engine

Table 5.5 Four-stroke trunk-type main engines

(a) *Cylinder head.* Disconnecting and removing cylinder head, cleaning all exposed parts, including piston crown and reassembling as before using all owner's supplied spares. Disconnecting and removing two in number air inlet valves and two in number exhaust valves, including exhaust valve cage with removable seats. Cleaning and decarbonizing valves, cages, and head as far as accessible, lightly hand-grinding valves for examination only of seating areas.
Clarifications: Work on the seats may be protracted so is excluded. Work could include changing seat inserts, machining, and grinding/lapping seats. This will require establishing and should be subject to work assessment.
(b) *Top overhaul.* Disconnect and remove one pair of crankcase doors, disconnect bottom end bearing fastenings. Disconnect and remove cylinder head, withdraw piston, remove piston rings, clean, calibrate, and reassemble as before using all owner's supplied spares.
(c) *Piston gudgeon pin.* Drawing out gudgeon pin from removed piston, cleaning all exposed parts, calibrating and recording, and reinstalling pin as before. Any spares to be owner's supply.

Continued

Table 5.5 Four-stroke trunk-type main engines—cont'd

Cylinder bore (mm)	(a) Man-hours per cylinder head	(b) Man-hours per cylinder	(c) Man-hours per piston pin
100	16	16	4
150	20	20	4
200	22	24	6
250	24	32	8
300	32	36	10
350	36	40	12
400	40	48	14
450	48	56	16
500	56	64	18
550	64	72	20
600	72	80	24

Note: For Vee bank engines add 20% per unit.

Table 5.6 Cylinder liners – 2

Withdrawing of cylinder liner, cleaning and painting exposed areas as far as accessible, installing of new, owner's supplied liner, or existing liner, complete with owner's supplied rubber seals and top joint. Attending hydrostatic test carried out by ship's staff.

Cylinder bore (mm)	Man-hours per cylinder
100	16
150	20
200	24
250	28
300	32
350	40
400	44
450	48
500	52
550	56
600	60

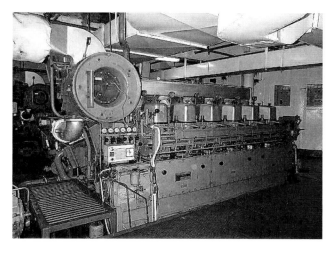

Figure 5.2 A ship's generator diesel engine

Table 5.7 Bearing survey – 2

Opening up for inspection, exposing both halves, cleaning, calibrating, and presenting for survey. On completion, reassembling as before.

	Man-hours per bearing	
Cylinder bore (mm)	Crank pin	Main
100	6	4
150	8	6
200	10	8
250	14	10
300	16	12
350	18	14
400	20	16
450	24	18
500	28	20
550	32	24
600	36	32

Note: For exposing top half of bearing shell only, charge 60% of above rates.

Table 5.8 Crankshaft deflections – 2

(a) Opening up crankcase door for access works and refitting on completion.
(b) Setting up deflection indicator gage, turning engine, using ship's powered turning gear, and recording observed readings. Removing equipment and closing up crankcase door on completion.

Cylinder bore (mm)	(a) Crankcase doors (man-hours per door)	(b) Deflections (man-hours per unit)
>100	4	4
100–200	4	6
200–300	4	8
300–400	6	12
400–500	8	16
500–600	8	20

Valves

Table 5.9 Overhauling valves, manually operated types

Opening up hand-operated, globe and gate valve for in-situ overhaul, by disconnecting and removing cover, spindle and gland, cleaning all exposed parts, hand-grinding of globe valve, light hand-scraping of gate valve, testing bedding, painting internal exposed areas and reassembling with new cover joint, and repacking gland with conventional soft packing.

For low-pressure valves, below 10 kg/cm^2, the tariff is the same as for sea valves. This also applies to the increases for location.

The following increases can be applied according to the pressure:

Pressure (kg/cm^2)	Increase over sea valve tariff (%)
10	100
16	120
20	140
40	160
60	180
80	200

Notes:
The valves are considered to be ship side sea valves or similar type inboard valves.
Pressure valves are inboard for water, oil, compressed air and the like.
Insulation renewal excluded.

Pressure testing in situ using ship's pump; additional 5 hours per valve.
Pressure testing in situ using contractor's pump; additional 7 hours per valve.
Removal and overhaul ashore in workshop; double the in-situ rate.
Metallic or special packings, owner's supply; additional 2 hours per valve.
Rates apply to hand-operated valves only.

Figure 5.3 A ballast system valve chest

Condensers

Table 5.10 Main condenser

Opening up inspection doors, cleaning sea water end boxes and tubes by air or water lance, test, and reclosing.

SHP	Man-hours
12,000	120
15,000	140
20,000	160
22,000	180
24,000	200
30,000	220

Note: Excluding stagings for access.

Heat exchangers

Table 5.11 Overhauling heat exchanger

(a) Disconnecting and removing end covers, cleaning water side end plates and water boxes and tubes by air or water lance, test, and reclosing.
(b) *Hydraulic testing.* Disconnecting and removing secondary side pipe works. Providing necessary blanks and installing. Filling with fresh water

and applying necessary hydraulic pressure test. Draining on completion, removing blanks and installing pipes as before.

Cooling water surface area (m^2)	Man-hours (a)	(b)
>3	16	8
5	20	12
10	24	12
15	32	12
20	36	16
30	40	16
50	44	16
100	56	16
150	64	20
200	80	20
250	88	20
300	96	24
350	108	24
400	120	32

Notes:
Applicable for standard tube type heat exchanger. For plate type, cleaning rates to be increased by 120%.
Including renewal of owner's supplied internal sacrificial anode on primary side.
Including painting with owner's supplied painting composition on primary side.
Excluding any repairs.
Excluding draining secondary side and associated cleaning works.
For ultrasonic cleaning, special considerations to apply.

Turbines

Table 5.12 Main steam turbines

Opening up for inspection, disconnecting flexible coupling and lifting up rotor, examining bearings, coupling and rotor, checking clearances, cleaning jointing surfaces, and reclosing.

	Man-hours	
SHP	HP turbine	LP turbine
12,000	180	210
15,000	220	250
20,000	250	280
22,000	260	300
24,000	270	310
26,000	280	320
32,000	350	400
36,000	430	480

Note: Removal of any control gear is not included. If applicable, increase rates by 30%.

Table 5.13 Flexible coupling

Disconnecting guard, opening up coupling, cleaning, presenting for survey and examination, measuring and recording clearances, closing up. Excluding any repairs, renewals, or realignments works.

SHP	Man-hours each
12,000	30
15,000	32
20,000	34
22,000	35
24,000	36
26,000	37
28,000	38
30,000	40
32,000	42
34,000	44
36,000	46

Note: Removals for access are excluded.

Table 5.14 Auxiliary steam turbines

Opening up for in-situ inspection, disconnecting flexible coupling and lifting up turbine rotor, examining bearings, coupling and rotor, checking clearances, cleaning jointing surfaces, and reclosing.

Turbo alternator turbine

kW	Man-hours each
400	170
700	180
1000	190
1500	200
2000	230
2500	270

Cargo pump turbine

Tonnes/hour (of pump)	Man-hours each
>1000	100
1500	110
2000	125
2500	145
3000	150
4000	165
5000	185

Feed pump turbine

SHP	Man-hours each
12,000	42
15,000	48
20,000	56
24,000	60
30,000	64
36,000	68

Notes:
Assume pump and turbine to be horizontal.
For vertical pump, increase by 10%.
Excluding dynamic balance checking.

Table 5.15 Water-tube boiler feed pumps (multi-stage type)

Disconnect and remove upper half of pump casing, disassemble internals and draw shaft. Clean, inspect, and calibrate all parts. Rebuild rotor, set clearances, and refit upper half of casing. All spares to be owner's supply.

SHP	Man-hours each
12,000	48
15,000	52
20,000	56
24,000	60
30,000	64
36,000	68

Notes:
The above figures apply to overhaul of pump only. If carried out in conjunction with overhaul of turbine also, then reduce figures by 10%.
All access works and insulation works excluded.

Table 5.16 Oil tanker cargo pumps

Disconnecting and removing top half of casing, releasing shaft flexible coupling from drive, slinging and removing impeller, shaft and wearing rings.

Withdrawing impeller, shaft sleeve, and bearings from shaft.

Cleaning all exposed parts, calibrating, and reporting. Reassembling as before using owner's supplied parts, jointing materials, and fastenings.

Tonnes/hour	Man-hours each
>1000	90
1500	100
2000	110
2500	125
3000	140
4000	150
5000	160

Notes:
Horizontal centrifugal single-stage type pumps.
For vertical pumps, increase figures by 15%.

Compressors

Table 5.17 Air compressor (two-stage reciprocating type)

Disconnecting and removing cylinder heads, releasing bottom end bearings, withdrawing pistons.
Opening up main bearings, including removing crankshaft on compressors with removable end plate.
Dismantling cylinder head air suction and delivery valves.
Cleaning all parts, calibrating, and reporting condition.
Reassembling all as before using owner's supplied spares as required.
Cleaning of attached air inter-cooler, assuming accessible.

Capacity (m^3/hour)	Man-hours per machine
10	50
20	55
50	60
100	70
200	80
300	90
400	100
500	120
600	150

Note: For three-stage compressors, charge rate to be increased by 150%.

Receivers

Table 5.18 Air receivers

Opening up manholes, cleaning internal spaces for inspection, painting internal areas, and closing manholes with owner's supplied jointing materials.

Capacity (m^3)	Man-hours per receiver
<1	16
5	20
10	24
15	28
20	32
30	36
40	40
50	48
60	52

Pumps

Table 5.19 Horizontal centrifugal-type pumps

Disconnecting and removing top half of casing, releasing shaft coupling from motor drive, slinging and removing impeller, shaft, and wearing rings.
Withdrawing impeller, shaft sleeve, and bearings from shaft.
Cleaning all exposed parts, calibrating, and reporting. Reassembling as before using owner's supplied parts, jointing materials, and fastenings.

Capacity (m^3/hour)	Man-hours per pump
5	20
10	24
20	32
50	34
100	40
200	48
300	52
500	56
750	60
1000	64
1500	72

Notes:
Assuming single-stage pump.
Assuming driven by an attached electric motor.
For multi-stage turbine-driven pumps, see the rates given for water-tube boiler feed pumps and assess accordingly.

Figure 5.4 A vertical electric-driven centrifugal water pump

Table 5.20 Reciprocating-type pumps, steam driven: (a) simplex; (b) duplex

Disconnecting and removing steam cylinder top cover, releasing steam piston, withdrawing, removing piston rings, cleaning, calibrating, and recording.

Disconnecting and removing slide valve cover, removing valves, cleaning, and presenting for survey.

Disconnecting and removing bucket cover, releasing bucket, withdrawing, removing bucket rings, cleaning, calibrating, and recording.

Opening up suction and delivery valve chest, removing valves and springs, cleaning, grinding, and presenting for survey.

Fully reassembling complete pump, renewing all jointing, and repacking glands with owner's supplied conventional soft packing.

Excluding all repairs and renewals.

	Man-hours	
Capacity of pump (m^3/hour)	(a) Simplex pump	(b) Duplex pump
50	64	112
100	80	140
200	104	182
300	120	210
400	144	252
500	160	280

Table 5.21 Reciprocating-type pumps, electric motor driven: (a) simplex; (b) duplex

Disconnecting and removing electric motor aside.
Disconnecting and removing bucket cover, releasing bucket, withdrawing, removing bucket rings, cleaning, calibrating, and recording.
Opening up suction and delivery valve chest, removing valves and springs, cleaning, grinding, and presenting for survey.
Fully reassembling complete pump, renewing all jointing, and repacking glands with owner's supplied conventional soft packing.
Reinstalling electric motor and making terminals.
Excluding all repairs and renewals.

Capacity of pump (m^3/hour)	Man-hours	
	(a) Simplex pump	(b) Duplex pump
5	32	56
10	40	72
20	48	80
30	56	96
40	64	112
50	72	120

Table 5.22 Gear-type pumps (helical and tooth)

Disconnecting and removing pump, opening up end covers, withdrawing gear units, cleaning, calibrating, recording clearances, and presenting for survey.
Fully reassembling pump, renewing all jointing, and repacking glands with owner's supplied packing or seals.

Capacity (m^3/hour)	Man-hours per pump
1	16
5	24
10	28
20	32
50	48
100	56
200	64
300	72
400	80
500	88

Table 5.23 Steering gear

Variable delivery constant-speed electro-hydraulic pumps.
Disconnecting pump and removing for in-situ overhaul. Opening up pump, full dismantling, cleaning, calibrating, and presenting for survey.
Full reassembling using owner's supplied spares and reinstalling in place.

Capacity (HP)	Man-hours per pump
3	24
5	32
8	36
10	42
15	48
20	56
25	64
30	72

Boilers (main and auxiliary)

Table 5.24 Cleaning of water-tube boilers

Opening up gas side of boiler, normal clean all fire side surfaces using HP fresh water. Removing drain plates from gas side of boiler to permit drainage of water to bilges and later refitment. Closing up gas sides as before.

Opening up water side of boiler by removing manhole doors from water drums, hosing down with fresh water, and reclosing.

| | Man-hours | | |
Vessel SHP	Single boiler	Two boilers	Auxiliary boiler
20,000	453	748	340
22,000	464	766	350
24,000	476	818	360
26,000	504	857	365

Notes:
Allowance made for 10 access doors, 10 hand-hole doors, and two manholes per boiler.
Special cleaning of super-heater tubes to be charged as additional.
Extra dirty boilers to be charged as additional.
Staging not included.
For hydraulic testing of boiler using fresh water, 10% additional charge. Excluding economizer and super-heater.

6 Electrical works

Table 6.1 Insulation resistance test on all main and auxiliary lighting and power circuits, and report

Vessel DWT	Man-hours
<5000	24
>5000	40
10,000	48
20,000	56
50,000	64
75,000	64
100,000	72

Figure 6.1 The main electrical switchboard in a machinery control room

Table 6.2 Switchboard

Cleaning behind switchboard, examining all connections, and retightening as necessary, reporting conditions.

Vessel DWT	Man-hours
<5000	32
>5000	40
10,000	50
20,000	75
30,000	100

Figure 6.2 A generator control panel in the main switchboard

Table 6.3 Electric motors – 1

Disconnecting motor from location, transporting motor ashore to workshop for servicing, and, on completion, returning on board, refitting in original position and reconnecting original cables.
Receiving motor in workshop, dismantling, removing rotor bearings, and cleaning all parts. Baking dry in oven, dip varnishing, and rebaking in oven. Reassembling all parts, fitting new standard type ball or roller bearings to rotor, and testing in workshop.

Continued

Table 6.3 Electric motors – 1—cont'd

Capacity (kW)	Man-hours
<3	16
5	20
10	24
15	27
20	27
30	32
40	34
50	39
60	48
75	54
100	60

Notes:
Excluding rebalancing of rotor.
These man-hours are for work on AC motors only, which are assumed to be single-speed, squirrel cage induction motors, three-phase, 380/440 volts, 50/60 Hz, 1440/1760 rpm, and with Class B insulation.
Excluding: staging for access to location, removals in way, cleaning in way, and cranage.

Table 6.4 Electric motors – 2

Disconnecting motor from location, transporting motor ashore to workshop for rewinding, and, on completion, returning on board, refitting in original position, and reconnecting original cables.

Receiving motor in workshop, dismantling, cutting out all stator coils, removing rotor bearings, and cleaning all parts. Forming new stator coils in copper wire, assembling using new insulation, and varnishing. Baking dry in oven, dip varnishing, and rebaking in oven. Reassembling all parts, fitting new standard type ball or roller bearings to rotor, and testing in workshop.

Capacity (kW)	Man-hours
<3	24
5	30
10	38
15	40
20	40
30	48
40	50
50	60
60	75
75	80
100	90

Notes:
Excluding rebalancing of rotor.
These man-hours are for work on AC motors only, which are assumed to be single-speed, squirrel cage induction motors, three-phase, 380/440 volts, 50/60 Hz, 1440/1760 rpm, and with Class B insulation.
Excluding: staging for access to location, removals in way, cleaning in way, and cranage.

Figure 6.3 A standard AC induction electric motor

Table 6.5 Electric motors for winch/windlass/crane – 1

Disconnecting motor from location, transporting motor ashore to workshop for servicing, and, on completion, returning on board, refitting in original position, and reconnecting original cables.

Receiving motor in workshop, dismantling, removing rotor bearings, and cleaning all parts. Baking dry in oven, dip varnishing, and rebaking in oven. Reassembling all parts, fitting new standard type ball or roller bearings to rotor, and testing in workshop.

Capacity (kW)	Man-hours
<20	110
30	135
40	150
50	160
60	200
75	220
100	250

Notes:

Excluding rebalancing of rotor.

These man-hours are for work on AC motors only and these are assumed to be triple-speed, three-phase, 380/440 volts, 50/60 Hz, double rotor/stator, squirrel cage induction motors with integral brake with Class B insulation.

Excluding: staging for access to location, removals in way, cleaning in way, and cranage.

Table 6.6 Electric motors for winch/windlass/crane – 2

Disconnecting motor from location, transporting motor ashore to workshop for rewinding, and, on completion, returning on board, refitting in original position, and reconnecting original cables.

Receiving motor in workshop, dismantling, cutting out all stator coils, removing rotor bearings, and cleaning all parts. Forming new stator coils, for both stators, in copper wire, assembling using new insulation, and varnishing. Baking dry in oven, dip varnishing, and rebaking in oven. Reassembling all parts, fitting new standard type ball or roller bearings to rotor, and testing in workshop.

Capacity (kW)	Man-hours
<20	160
30	200
40	220
50	240
60	300
75	320
100	360

Notes:

These man-hours are for work on AC motors only, which are assumed to be triple-speed, three-phase, 380/440 volts, 50/60 Hz, double rotor/stator, squirrel cage induction motors with integral brake with Class B insulation.

Excluding: staging for access to location, removals in way, cleaning in way, and cranage.

Table 6.7 Electric generators

Disconnecting and removing rotor only from AC alternator ashore to workshop, full cleaning, baking in oven, drying, varnishing, rebaking in oven, testing, reassembling, and reconnecting in place as before in original position on board and reconnecting original cables.

KVA	Man-hours
<50	75
51–100	90
101–200	100
300	140
400	160
500	175
600	190
750	200
1000	220

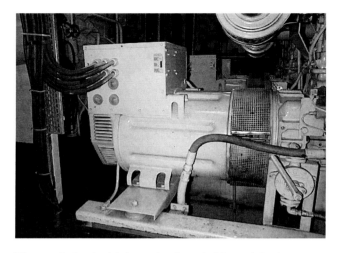

Figure 6.4 A ship's main diesel-driven AC alternator

Table 6.8 Installation of electric cables – 1 (man-hours for installations per 100 meters of unbraided, flexible, multi-core, rubber-insulated cable)

Including:
- Handling and installing in place numbers of lengths as indicated of electric cable.
- Stripping back cable and insulation and preparing for connection.
- Connecting to existing junction boxes in existing cable tray with new cable ties.
- Man-hours shown are for installation of exposed cables up to heights of 3 meters on exposed flat surfaces in existing cable trays.

Excluding:
- Material costs. These figures show man-hour charges only.
- Any removals of existing or old cable. These man-hours are for new installations only.
- Installation of scaffolding and any access work. These to be covered in separate sections.

Area (mm^2)	2 core	3 core	4 core	5 core	6 core	7 core	9 core	12 core
0.5	21	25	30	30	34	37	—	—
1	23	27	33	34	38	41	—	—
1.5	25	33	49	57	65	82	—	—
2.5	43	49	54	66	78	90	106	122
4	48	53	61	72	89	106	129	150
6	54	61	66	79	—	115	—	—
10	61	72	77	—	—	—	—	—

Continued

Table 6.8 Installation of electric cables – 1 (man-hours for installations per 100 meters of unbraided, flexible, multi-core, rubber-insulated cable)—cont'd

Notes:

Man-hours shown are for the installation of a single length of cable.

For the installation of a parallel, second length of electric cable, reduce by 15%.

For the installation of a parallel, third length of electric cable, reduce by 25%.

For the installation of a parallel, fourth length of electric cable, reduce by 30%.

For the installation of a parallel, fifth and subsequent length of electric cable, reduce by 35%.

For additional height, increase tariff as follows:

3–5 meters, increase by 5%.

5–8 meters, increase by 10%.

8–12 meters, increase by 15%.

Table 6.9 Installation of electric cables – 2 (man-hours for installations per 100 meters of rubber-insulated, or similar, braided flexible cable, braided in bronze or steel, basket weave)

Including:
- Handling and installing in place numbers of lengths as indicated of electric cable.
- Stripping back cable and insulation and preparing for connection.
- Connecting to existing junction boxes in existing cable tray with new cable ties.
- Man-hours shown are for installation of exposed cables up to heights of 3 meters on exposed flat surfaces in existing cable trays.

Excluding:
- Material costs. These figures show man-hours charges only.
- Any removals of existing or old cable. These man-hours are for new installations only.
- Installation of scaffolding and any access work. These to be covered in separate sections.

Area (mm^2)	2 core	3 core	4 core	5 core	6 core
1	28	36	40	45	51
1.5	29	39	51		
2.5	46	59	67		
4	53	67	74		
6	74	86	92		
10	82	90	94		
16	86	93	99		
25	90	95	109		

Continued

Table 6.9 Installation of electric cables – 2 (man-hours for installations per 100 meters of rubber-insulated, or similar, braided flexible cable, braided in bronze or steel, basket weave)—cont'd

Area (mm^2)	2 core	3 core	4 core	5 core	6 core
35	95	99	120		
50	99	112	141		
70	104	132	164		
95	109	141	180		
120	128	167	205		
150	132	176	218		
185	154	205	256		
240	196	261	326		
300	254	352	440		

Notes:

Man-hours shown are for the installation of a single length of cable.

For the installation of a parallel, second length of electric cable, reduce by 15%.

For the installation of a parallel, third length of electric cable, reduce by 25%.

For the installation of a parallel, fourth length of electric cable, reduce by 30%.

For the installation of a parallel, fifth and subsequent length of electric cable, reduce by 35%.

For additional height, increase tariff as follows:

3–5 meters, increase by 5%.

5–8 meters, increase by 10%.

8–12 meters, increase by 15%.

Figure 6.5 Grouping of electric cables on a cable tray

Table 6.10 Installation of electric cables – 3 (man-hours for installations per 100 meters (rubber-insulated, or similar, braided flexible, cable, braided in bronze or steel, basket weave; single-core cable, for use with multi-runs of higher cable sizes)

Including:
- Handling and installing in place numbers of lengths as indicated of electric cable.
- Stripping back cable and insulation and preparing for connection.
- Connecting to existing junction boxes in existing cable tray with new cable ties.
- Man-hours shown are for installation of exposed cables up to heights of 3 meters on exposed flat surfaces in existing cable trays.

Continued

Table 6.10 Installation of electric cables – 3 (man-hours for installations per 100 meters (rubber-insulated, or similar, braided flexible, cable, braided in bronze or steel, basket weave; single-core cable, for use with multi-runs of higher cable sizes)—cont'd

Excluding:
- Material costs. These figures show man-hours charges only.
- Any removals of existing or old cable. These man-hours are for new installations only.
- Installation of scaffolding and any access work. These to be covered in separate sections.

Area (mm^2)	1 wire	2 wires	3 wires	4 wires	5 wires
50	80	88	102	122	157
70	92	102	117	140	182
95	108	119	137	163	213
120	125	137	157	189	246
150	132	145	166	200	260
185	148	163	188	225	293
240	174	191	220	263	342
300	231	254	293	351	456
380	281	310	356	427	556

Notes:
Man-hours shown are for the installation of single runs of cable without joins.
For additional height, increase tariff as follows:
3–5 meters, increase by 5%.
5–8 meters, increase by 10%.
8–12 meters, increase by 15%.

Table 6.11 Installation of electric cable tray (man-hours for installations of cable tray per meter; perforated steel cable tray, including brackets and fastenings)

Including:
- Handling and installing in place numbers of lengths as indicated of electric cable tray.
- Man-hours shown are for installation of exposed cable tray up to heights of 3 meters on exposed flat surfaces.

Excluding:
- Material costs. These figures show man-hour charges only.
- Any removals of existing or old cable tray. These man-hours are for new installations only.
- Installation of scaffolding and any access work. These to be covered in separate sections.

Size (mm)	Man-hours per meter
75	0.95
100	1.05
150	1.25
225	1.50
300	1.75
Bends	Each bend to be rated at 3 times that of 'per meter'

Notes:
Man-hours shown are for the installation of a single length of cable tray.
For additional height, increase tariff as follows:
3–5 meters, increase by 5%.
5–8 meters, increase by 10%.
8–12 meters, increase by 15%.

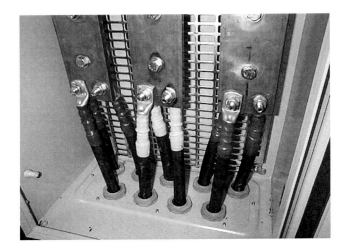

Figure 6.6 Part of a distribution panel with cable attachments

Table 6.12 Installation of electric cable conduit (man-hours for installations of electric cable conduit per meter; galvanized steel conduit, including brackets and fastenings)

Including:
- Handling and installing in place numbers of lengths as indicated of electric cable conduit.
- Man-hours shown are for installation of exposed cable conduit up to heights of 3 meters on exposed flat surfaces.

Excluding:
- Material costs. These figures show man-hour charges only.
- Any removals of existing or old cable conduit. These man-hours are for new installations only.
- Installation of scaffolding and any access work. These to be covered in separate sections.

Size (mm)	Man-hours per meter
20	0.72
25	0.78
32	1.14

Notes:
Man-hours shown are for the installation of a single length of cable conduit.
For additional height, increase tariff as follows:
3–5 meters, increase by 5%.
5–8 meters, increase by 10%.
8–12 meters, increase by 15%.

7 General works

Table 7.1 General cleaning

(a) Berthing vessel alongside special tank cleaning berth.
(b) Receiving of bilge water or slops into shore facility using ship's pumps.
(c) Removing sludge deposits from tanks and disposal ashore.

(a) Man-hours (minimum charge)	(b) Man-hours per 20 tonnes (minimum)	(c) Man-hours per 10 tonnes (minimum)
150	1.5	105

Notes:
(a) This rate may vary depending upon shipyard. An hourly rate will apply with a minimum charge being levied, as shown in the table above.
(b) The rate for collection of bilge water or slops will depend upon the receiving facility and the rate levied for (a). For collection by road tankers, a separate quote should be requested.

Table 7.2 Tank cleaning

(a) Removal of tank manhole cover for access and refitting with new cover joint.
(b) Removing dirt and debris per cubic meter.
(c) Hand-cleaning of bilge areas or inside tanks per 10 square meters.
(d) Hand-scraping of internal steel areas per 10 square meters.

	Man-hours			
Type of tank	(a)	(b)	(c)	(d)
Fresh water	6	0.70	1.25	1.0
Ballast water	6	6.3	1.60	1.70
Fuel oil (MGO)	6	10.5	4.25	—

Table 7.3 Tank painting

Applying owner's conventional paint, per coat, per square meter by brush after completion of tank cleaning works.

Type of tank	Man-hours
Fresh water	0.1
Ballast water	0.1
Fuel oil (MGO)	0.1

Table 7.4 Tank testing

(a) Tank testing by low-pressure compressed air, per tonne capacity.
(b) Tank testing by filling with sea water, per tonne capacity.

| | Man-hours | |
Tank capacity (m³)	(a)	(b)
<5	0.25	0.32
5–20	0.20	0.25
20–50	0.16	0.20
50–100	0.14	0.16

Special notes on quotations:

- Obtain a copy of the ship repair contractor's standard tariff rates.
- Request the ship repair contractor to agree that extra work will be priced in accordance with produced standard tariffs, or other agreed rates.
- Ensure that conditions of contract are agreed before placing of contract. If not, then it will be assumed that the shipyard's standard conditions apply, which may not always be suitable to the shipowner.

8 Planning charts

The following is not necessarily required by shipowners' superintendents, but may prove useful to give an indication as to a method of determining the timescale and daily loadings for carrying out the repairs.

In forward planning and scheduling it is imperative that the planned timescales for repair periods are adhered to strictly in order to avoid knock-on effect delays. A ship repair yard therefore must be aware, well in advance, of the total workload and resources needed to complete each job. This is where the man-hour totals for each trade are required and, very importantly, the work rate of each trade.

The graphs shown in this section have been compiled from historical data by shipyard workload planners and are actual graphs derived, and used by, a large international ship repair yard to assist the forward planning of the yard. The yard's planners must ensure that sufficient resources are available to carry out the workload, looking up to 3 months ahead, and this is their method of doing so. Using this process the planners can arrange for the necessary resources to be available well ahead of the scheduled repair period and have these available on arrival of the vessel.

Using a prepared ship repair specification, a planner will carry out an analysis of the work and produce a critical path. This critical path determines the timescale of the repair period, so any way in which the timescale of a job within the critical path may be reduced will reduce the overall timescale. Additional resources will be used on these jobs to ensure their earlier completion; so then this is the way in which the total timescale is reduced.

Using the foregoing tables in this book, the estimator can forecast the total number of man-hours per trade for the specified work. Knowing the yard's resources, the next job is to develop the daily work rate for each trade.

A graph of the work rate for each trade is available with the yard's planners and, using the graphs, the planner can estimate the timescale to complete the known works.

In carrying out a planned repair period, the planner will consider certain aspects of priority.

As an example, consider a vessel entering the dry dock and a number of trades having planned work in the dry dock area. A very high work rate is necessary to complete any work that prevents other trades from carrying out their work. Into this category come the hull treatment workers. These are the first workers on the external areas of the vessel. This trade will hand-scrape the hull free from sea growth and then carry out high-pressure jet washing of the hull. Preparations will then be made for this trade to continue with grit-blasting to clean the hull and then apply the first coat of primer paint. Once this high-activity area has been completed, the work rate of the hull treatment trade may be reduced to make way for other trades to carry out their external work. The hull treatment trade workload may now be reduced, and certain of these workers released to other high-activity areas.

During this period, very few other trade workers will be able to work in the same vicinity, so the planners assign these to other areas.

The trade graph will indicate the high work rate of the hull treatment workers initially on the hull and then show the tapering off.

All trades are considered in a similar manner and graphs drawn from historical data until the work rate of each trade can be predicted.

The graphs have been drawn up indicating the trend of work rate of the individual trades and are used to determine the timescale of the repair period.

Conflicts always occur in repair periods. As noted with the hull treatment, no other trade can work during this period, so this constitutes a conflict in this area. There are many conflicts between trades and also within trades, causing delays in starting jobs and continuing jobs. The jobs on the critical path generally are given a higher priority than other jobs by the overall coordinator of the work.

The following example describes the method of using the workload graphs.

As an example, take the marine fitter graph. The estimator/planner will have determined the total man-hours for the complete specified works so will have a grand total.

Knowing the available resources at the yard, the maximum amount/number allocated to a ship will be known, e.g. 10 men. The percentage work is an estimated total, e.g. 1000 man-hours. Each man may be assigned to work 10 hours per shift. So the logical time to complete the works will be:

1000 man-hours/10 men × 10 hours per shift = 10 shifts

When this is determined, a decision will be made on whether this time is excessive and, if so, additional resources will be assigned. If not, then the work will continue as planned.

Ten men × 10 hours per shift = 100 man-hours per shift should be 10% of the work per shift. Carrying out this constant work rate would produce a straight-line graph at 45° where the slope would be 'y' = 'x'.

However, this does not happen and is shown from the marine fitter graph as follows:

Each work shift comprises 100 man-hours.
The first shift's work will complete 5% of the work.
The second shift's work will continue up to 14% of the work, an increase of 9%.
The third shift's work will continue up to 31% of the work, an increase of 17%.
The fourth shift's work will continue up to 43% of the work, an increase of 12%.

The fifth shift's work will continue up to 55% of the work, an increase of 12%.

The sixth shift's work will continue up to 67% of the work, an increase of 12%.

The seventh shift's work will continue up to 81% of the work, an increase of 14%.

The eighth shift's work will continue up to 88% of the work, an increase of 7%.

The ninth shift's work will continue up to 95% of the work, an increase of 7%.

The tenth shift's work will continue up to 100% of the work, an increase of 5%.

This indicates the varying degrees of output for the same man-hour input. This is caused by the types of conflict shown in the hull treatment workers, where other trades must allow them sole access to certain areas.

The trade supervisors, together with the planners, may increase or decrease the quantity of workers in certain areas to allow smooth running of the work.

This is one use of the graphs. Another is where a vessel must be completed within a certain timescale and the graphs are used to indicate the numbers of workers per trade that must be used in order to meet the target date. Knowing this, if the yard does not possess the full resources themselves, then the planners can ascertain the number of subcontracted laborers that are required to complete the total work schedule.

In this instance the timescale will be a known entity, so then it will be established what are the exact number of man-hours per shift per trade to complete the work in accordance with the work rate of the trade graph. A histogram would then be drawn indicating the number of men per trade per shift, and these resources would be allocated well in advance of the arrival date of the vessel.

To illustrate this, again take the case of the marine fitters having a workload of 1000 man-hours and an exact time of 10 shifts being allocated to complete the job:

The first shift's work will complete 5% of the work = 50 man-hours.
The second shift's work will complete 9% = 90 man-hours.
The third shift's work will complete 17% = 170 man-hours.
The fourth shift's work will complete 12% = 120 man-hours.
The fifth shift's work will complete 12% = 120 man-hours.
The sixth shift's work will co 12% = 120 man-hours.
The seventh shift's work will co 14% = 140 man-hours.
The eighth shift's work will complete 7% = 70 man-hours.
The ninth shift's work will complete 7% = 70 man-hours.
The tenth shift's work will complete 5% = 50 man-hours.
Total man-hours = 1000.
Total shifts = 10.

The manpower input is variable in accordance with the graph work rate loading.

The workers can therefore be assigned against the shift man-hour totals necessary to complete each shift's total workload.

The following histogram indicates the calculated shift totals shown above and the planners and repair coordinators can assign the daily trade manpowers accordingly.

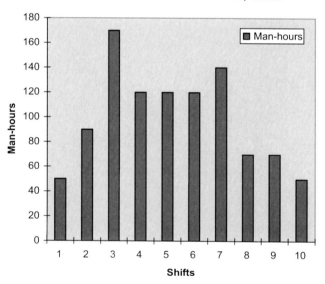

Sample graph loadings for major trades in ship repairing

(It should be noted that these sample graphs are actual loadings that were used by a certain major ship repair yard from figures compiled from production feedback over a number of years. These graphs were then used by the commercial division planners to predict the required manpower resources for up to 3 months ahead.)

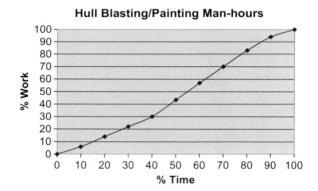

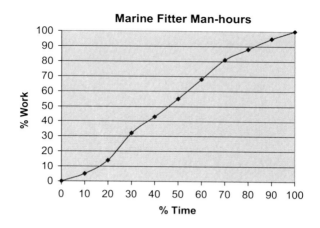

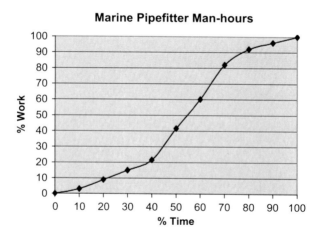

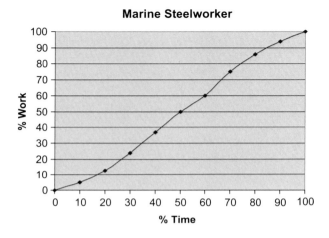

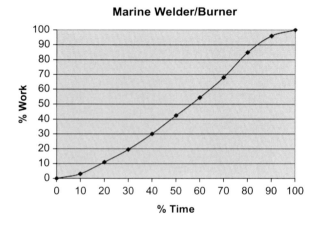

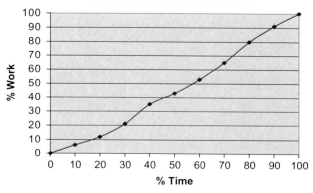

Index

Page references followed by "f" indicate figure, and by "t" indicate table.

A

Air compressors, 74
  three-stage compressors, 74t
Air receivers, 75
Alternator, 92f
Aluminum anodes weight determination, 28–29
Anchors, 35
Anodes, 25–29
  aluminum anodes weight determination, 28–29
  zinc anodes weight determination, 26–27

B

Bearing, 56t, 62t
Berth preparation, 5
Bilge water removal, 103t
Boilers (main and auxiliary), 82
  cleaning of, 82t
  feed pumps, 72t
Boot-top area, 14–15
Bulwarks area, 14–15

C

Cables, 35
Capacity of material, 26–27
Cargo pump turbine, 70t–71t
Chain lockers, 36
Condensers, 66
Copper pipes, 49
Crack detection, 23–24
Crankshaft deflections, 57t, 63t

Current, 26–27
Current density of material, 26–27
Cylinder bore, 54t–57t, 59t–63t
Cylinder cover, 55–57
Cylinder head, 59t–60t
Cylinder liners, 55t, 60t–61t

D

Deck plates, 45f
Design life, 26–27
Diesel engine overhauls, 53–63
Dock rent per day, 7–8
Docking, 6
Docking plugs, 30
Dry berth *see* Berth preparation
Drydocking, 5–37

E

Economical climate, impact on docking, 6
Electric cables, 93t–96t, 97f, 97t–99t, 100f, 101t
Electric motor, 79t, 85t–87t, 88f, 89t–90t
Electrical works, 83–101
Electro-hydraulic pumps, 81t

F

Feed pump turbine, 70t–71t
Fenders, 34
Flat steel plate, 41–42
Flexible coupling, 69t
Four-stroke diesel engine, 58–63

G

Galvanizing, 49
Gear-type pumps, 80t
General works, 103–105
Generator, 61f, 91t
Gland seal, 24t
Grit blast, 10–11

H

Heat exchangers, 60t–61t
Histogram, 111–113
Hollow fenders, 34t
Horizontal centrifugal-type pumps, 76t
Hot dip galvanizing, 49
Hull painting, 11–15
  notes for, 13
  paint area determination, 14–15
  painting compositions supply, 13–14

Hull preparation, 9–11
 special notes for, 9–11
Hull treatment, 108–109
Hydraulic testing, 66t–67t

I
Insulation resistance test, 83t

J
Jet wash, 11

K
Keel block, 5t, 6f

L
Liners *see* Cylinder liners

M
Marine fitter graph *see* Planning charts
Mechanical works, 53–82

O
Oil tanker cargo pumps, 73t
Overhauls, 53–63

P
Pipe works, 47–52
Piston gudgeon pin, 59t–60t
Planning charts, 107–117
 sample graph loadings, 114–117
Pre-arrival plan, 10–11, 111–112
Pre-planning *see* Planning charts
Pressure testing, 64t–65t
Propeller works, 18–20
 propeller polishing, 20t
Pumps, 76–81

R
Ready galvanized pipes, 49
Reciprocating-type pumps
 electric motor driven, 77f, 79t
 steam driven, 78t
Rewinding, 86t–87t, 90t
Rudder works, 16–17

S
Sea chests, 30
Sea valves, 31t–32t

Shifting of blocks after docking vessel, 5t
Ship plate renewals, 41–42
Ship side storm valves, 33t
Side block, 5
Simplex-type seal, 24t
Staging, 37
Steam turbines, 68t, 70t–71t
Steel works, 39–46
    flat steel plate, 41–42
    renewals, 43t–44t, 45–46
    steel angles, 42–43
Steering gear, 81t
Strainers, 30t
Switchboard, 84f, 84t, 85f

T
Tailshaft works, 21–24
    crack detection, 23–24
    tailshaft removal, for survey, 22–24
Tank cleaning, 104t
Tank painting, 104t
Tank testing, 105t
Top overhaul, 54t, 59t–60t
Topsides area, 14–15
Turbines, 68–73
Turbo alternator turbine, 70t–71t
Two-stroke diesel engine, 53–57

U
Underwater area, 14–15
Undocking, 6

V
Valves, 31–33, 64–65

W
Water-tube boiler
    cleaning of, 82t
    feed pumps, 72t
Welding, 34t
Work rate graph *see* Planning charts

Z
Zinc anodes weight determination, 26–27